YOUR KNOWLEDGE HAS VALUE

- We will publish your bachelor's and
  master's thesis, essays and papers

- Your own eBook and book -
  sold worldwide in all relevant shops

- Earn money with each sale

Upload your text at www.GRIN.com
and publish for free

Bibliographic information published by the German National Library:

The German National Library lists this publication in the National Bibliography; detailed bibliographic data are available on the Internet at http://dnb.dnb.de .

Imprint:

Copyright © 2014 GRIN Verlag, Open Publishing GmbH
Print and binding: Books on Demand GmbH, Norderstedt Germany
ISBN: 978-3-668-03950-6

This book at GRIN:

http://www.grin.com/en/e-book/304160/developing-high-finesse-cavities-for-phase-contrast-electron-microscopy

Kevin Li, Andrew Lee, Brian Lai

Developing High-Finesse Cavities for Phase Contrast Electron Microscopy

GRIN Publishing

Developing High-Finesse Cavities for Phase Contrast Electron Microscopy

Kevin Li, Andrew Lee, Brian Lai

Siemens 2014 Team Competition

September 2014

Abstract:

The transmission electron microscope is an indispensable tool in science, with applications across medicine, materials science, and geology, among others. However, it is limited in its ability to operate with Zernike phase contrast, a technology commonplace in light microscopy. Zernike phase contrast can be obtained, but only by using carbon-film phase plates or similar methods, all of which are short-lived. Electrons moving close to the speed of light cause damage as they bombard the phase plates. The phase plates need to be replaced frequently, which introduces inconsistencies due to variations between the plates as they are replaced. The purpose of this paper is to demonstrate the plausibility of utilizing ponderomotive forces within an optical cavity to achieve phase contrast, creating a laser-based phase plate, thereby replacing the carbon films and eliminating swapping. We approach this problem by using a Fabry-Perot to concentrate the laser power to be able to achieve the necessary electron phase shift with conventional $CO2$ lasers. We demonstrate a cavity with finesse of ~24000 and numerical aperture of ~.016, and calculate the laser power needed to be supplied to be ~19W, well within the state of art. These results demonstrate the practicality of laser-based electron microscope phase plates.

1. Introduction:

The electron microscope is an indispensable tool that allows scientists to magnify specimens up to ten million times. This ability is driving research in areas such as medicine, materials science, and geology. For example, it allows us to very closely observe bacteria for developing treatment and cures for elusive diseases. However, the current model for transmission electron microscopes is limited: many soft-matter specimens (such as biological specimens) do not absorb electrons well and thus produce images with poor contrast. With the development of phase contrast microscopy, the field of electron microscopy could take a giant leap forward. The current implementations, however, still leave much to improve; there does not exist an enduring apparatus to produce the desired effect in a reproducible manner. Here, we demonstrate a Fabry-Perot cavity with high finesse, a crucial component for constructing a phase contrast electron microscope that overcomes present limitations and produces reproducible images.

For this work, we combine techniques from the fields of laser optics and electron microscopy in a new way in order to improve upon previous methods of phase contrast electron microscopy. Phase contrast was first developed in light microscopes in the 1930s. Prior to then, optical microscopes produced images with little contrast for certain specimens. This posed many issues, especially for biologists, as images produced with low contrast showed little definition of the structure of the specimen. In the early 1930s, Frits Zernike addressed this issue by developing the technique of phase contrast microscopy [1]. This provided a means of revealing the more intricate details of the structure of the specimen being observed by obtaining contrast from specimen that do not absorb light. While earlier microscopes only showed contrast for objects that do absorb light-called amplitude objects-Zernike realized that even an object that did not absorb light may still shift the phase of light waves. Alas, this is invisible to the eye. Zernike then pioneered phase contrast, a method that could convert phase shifts into a visible image. This invention improved contrast and the functionality of optical microscopes by making use of an otherwise unused resource.

The principle behind phase contrast microscopy is that part of the light passing through the specimen is diffracted, while another part is left unperturbed. In a phase contrast microscope, the non-diffracted light is passed through a phase plate, located in the focal plane of the object lens. This plate phase-shifts the undiffracted light by $\pi/2$ radians. The diffracted and undiffracted light then propagate into the image plane, where they interfere. The phase shift produced by the phase plate alters this interference and produces phase contrast. The structures of the specimen are thus brought into the foreground and previously-unresolvable details in structure are enhanced (see Fig. 1). This technique has since become an integral mechanism in light microscopes, but has not been attempted in transmission electron microscopes until recently.

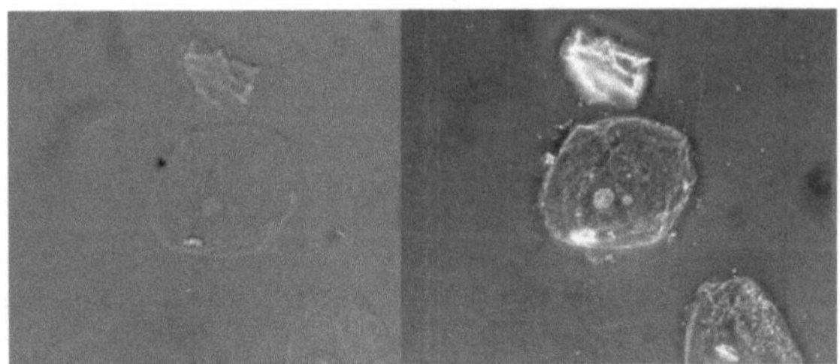

Fig. 1: an image obtained using traditional methods of bright field microscopy (left) and an image obtained using phase contrast microscopy (right) [2]

Unlike optical microscopes, electron microscopes regularly produce images with little contrast because biological specimens do not absorb electrons well. In attempt of realizing the

Zernike phase shift in electron microscopy, numerous methods were proposed and tested. These include so-called einzel lenses (an electrostatic model), thin magnetic bars, and carbon films. Of these, both the einzel lenses and thin magnetic bars are susceptible to becoming electrostatically charged when hit by the electron beam, which in turn distorts the image produced. Furthermore, the implementation of the carbon film (a microfabricated film) induces electron loss and consequently, diminishes performance [3,4,5]. Although the carbon phase plate could effectively produce the Zernike phase shift, it's efficacy considerably declines as the device ages, forcing the plate to be replaced every thirty minutes. The root of the issue lies in that the carbon film cannot withstand the continuous bombardment of electrons moving at up to two-thirds the speed of light. This presents challenges when looking at specimens for longer periods of time. Because each phase plate differs structurally from any another due to inherent engineering imperfections, each image produced deviates from those produced by other carbon plates. This introduces an extra variable caused by the carbon film, thus leading to incongruities between images.

We sought to improve current methods of achieving the Zernike phase shift by employing a high power Nd:YAG laser (Neodymium-doped yttrium aluminium garnet laser), in lieu of a carbon plate. Our goal was to avoid the limited lifespan associated with carbon plates and similar technology. We can use the beam to the same effect as the carbon film, eliminating the need for continuous replacement of an expensive and inconsistent device.

Our proposed laser phase plate works as follows: A stream of photons from the laser collides with electrons shot out from the electron gun from within the electron microscope; this invokes a phenomenon known as the ponderomotive force. As a direct result of the collision between the two particles, the wavelength of the electron increases and thus provides the phase shift necessary for phase contrast electron microscopy. To this purpose, a particular Fabry-Perot cavity with a high finesse and a relatively high numerical aperture must be constructed in order to attain sufficient power and harvest more light within the cavity.

In this article, we discuss the usage of a Fabry-Perot cavity in order to create a tightly focused laser cavity as outlined in the design of the laser-based Zernike phase plate theorized by Holger Müller and Bob Glaeser at the University of California, Berkeley, by utilizing mirrors of extremely high reflectivity to create a Fabry-Perot cavity with uniquely high finesse and numerical aperture (NA) [6]. We have collected evidence that shows that laser-based phase contrast is possible.

2. Materials and Methods:

In transmission electron microscopes, objects are magnified based on the same principles that govern how light microscopes work. Instead of photons, electrons are used. They are manipulated using electromagnetic lenses, which behave much like ordinary lenses to photons, to pass through a specimen and produce an image on the viewing screen.

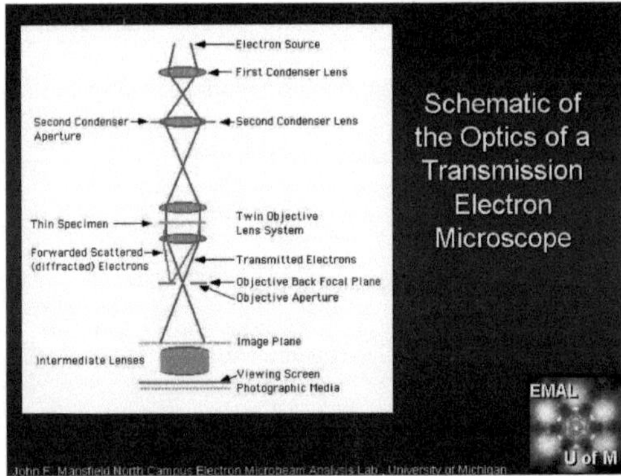

Fig. 2: Schematic of a Transmission Electron Microscope [7]

In the design of an electron phase contrast microscope, electrons are shot first through a sample and then an objective lens (see Fig 2). From here, our design differs from the traditional one in that the electrons continue into a small aperture which leads them into our optical cavity instead of a Zernike phase plate. In order to obtain a cavity where electrons can be retarded sufficiently strongly, the laser beam's intensity is enhanced by resonance inside said cavity. The cavity consists of two dielectric spherical output coupler mirrors facing each other. The semi-transparent backsides of these mirrors let incoming laser light in, and a small percentage of the light resonating inside out. The purpose of this cavity is to provide an intense spot of laser radiation which retards, or slows down, the electrons through the ponderomotive force so that they come out phase shifted.

Once the light beam enters the cavity, electrons are shot through an aperture on the side of the cavity. From here, the electrons are affected by a ponderomotive force from the electric field of the photons in the cavity. The ponderomotive potential energy, U_p, is given by

$$U_p = e^2 E_a^2 / 4m \, \omega_0^2 \text{, (1)}$$

where e is the elementary charge, m the mass of the electron, ω is the angular frequency of the oscillating electric field. The phase shift on the electron is then given by $U_p/\hbar$ integrated over the trajectory of the electron.

We sought to develop a Fabry-Perot cavity with a high finesse and relatively high numerical aperture, two characteristics of a cavity that are in an inverse relationship. We achieved this by using supermirrors, highly reflective mirrors with many layers of stacked dielectric coating. We worked with mirrors with a reflectivity of 99.99% at our wavelength of

1.064 microns. Our mirrors have a focal length of 1/2" ,so to achieve the concentric case for the desired finesse and numerical aperture, we made the distance between the mirrors approach 1".

As we adjust the mirror length with a micrometer, the finesse, numerical aperture, and coupling efficiency should increase as we approach a distance L of 1" and then disappear rapidly after we exceed 1", as we come in with a beam of large NA (see Eq. 2 below). We achieved this by placing a microscope objective in front of our cavity (see Fig. 5), allowing the beam to enter from a steep angle of incidence.

In obtaining a cavity with a high finesse and high numerical aperture, precision in the distance and angle between the mirrors was imperative for collecting data. Because of many disturbances in the set up including thermal expansion of the mirrors, collection of dust, and human disturbances of the apparatus, we decided to implement the Pound-Drever-Hall (PDH) method to lock the cavity. The PDH method in short creates a feedback loop from the back reflect of the cavity into the laser so that the laser can adjust the beam's wavelength to resonate in the Fabry-Perot cavity (see Fig. 3).

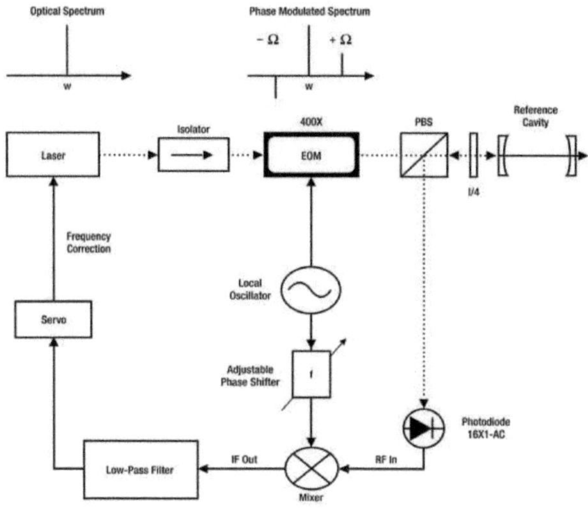

Fig. 3: A Schematic for the Pound-Drever-Hall Lock from the Newport website [8]

In our setup, we placed a quarter waveplate in the set up before the cavity so that upon entering the cavity the beam would be circularly polarized. Then, the back reflection would pass through the quarter waveplate on its way back and become polarized perpendicular to the original beam. We then place a polarizing beam splitter so that the back reflect is directed into a photodiode which then produces an error signal, a representation of the misalignment of the laser. This error signal (see Fig. 4) is what allows the laser to lock itself to the cavity.

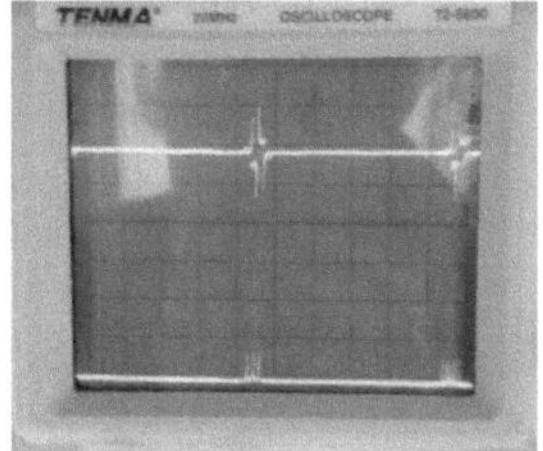

Fig. 4: The error signal on our oscilloscope. The error signal is represented by the top line.

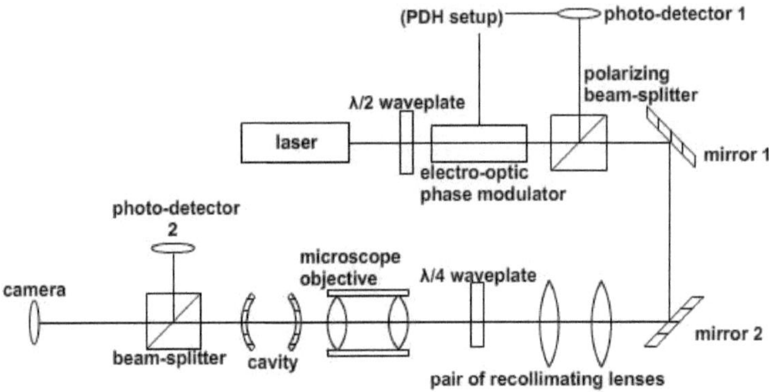

Fig. 5: The schematic of the optical setup on our table

We built the setup on the optical table, which holds all the optical and electrical components, by adjusting mirrors and lenses, along with wiring and electronics. Among those who helped us are our mentor, and a post-doctorate working in our professor's group. They helped answer our questions in times of difficulty or confusion. Furthermore, there are also several undergraduate and graduate students working in the neighboring labs in our professor's research group who were able to assist us with some of our work.

3. Theoretical Predictions:

Our work in laser cavities was mainly about finding a cavity that suited our needs for phase contrast electron microscopy. As such, the mirrors we used were specifically designed to optimize reflectivity, and also keep a relatively high numerical aperture. The cavity produced yields unique characteristics tailored to our needs - a very high finesse and moderate NA.

3.1. Numerical Aperture:

The acceptance angle and numerical aperture (NA) can be predicted from the equations governing Gaussian laser beams [9]. We start by knowing that the radius of curvature of beam

$R_z = z(1 + (\frac{z_R}{z})^2)$, where z is the distance from the waist, or narrowest point of the beam, and z_R is the Rayleigh range, or distance from waist where half the thickness of the beam is $w_0\sqrt{2}$, where w_0 is the width of the beam waist (see Fig. 6).

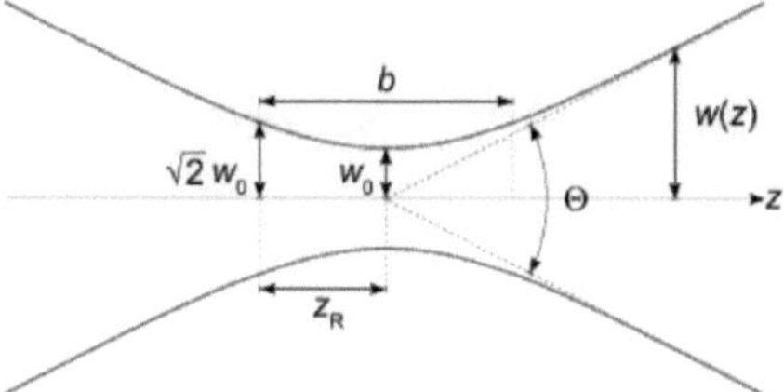

Fig. 6: A diagram of the various variables of a Gaussian beam. This image shows the

beam at it's narrowest point [9]

If we let R_z in our equation be the radius of our mirrors, R_M, we get $R_z = R_M = R(L/2)$, where L is the length between our mirrors (see Fig. 7).

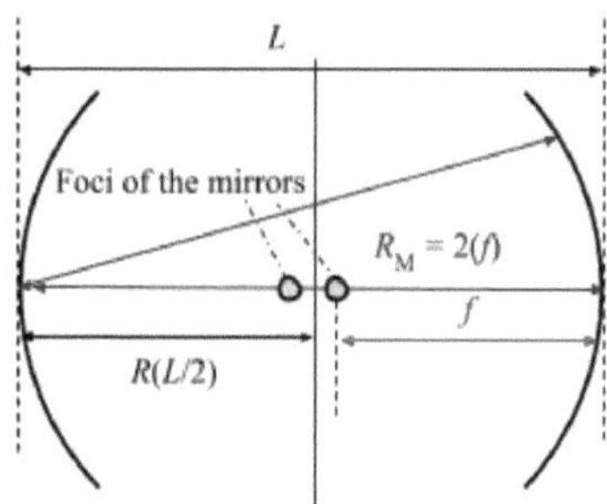

Fig. 7: A diagram showing the relationship between L, R_M, $R(L/2)$, and the focal distance (f)

This implies that $R(L/2) \approx R_M = \frac{L}{2}(1 + (2\frac{z_R}{L})^2)$. Solving for z_R, we get $z_R = \frac{L}{2}\sqrt{\frac{2}{L}(R_M - 1)}$. We also

know that the Rayleigh range $z_R = \frac{\pi w_0}{\lambda}$. Substituting this value for z_R, we get $\frac{L}{2}\sqrt{\frac{2}{L}(R_M - 1)} = \frac{\pi w_0}{\lambda}$.

Then, by solving for w_0, we get $w_0 = \sqrt{\frac{L\lambda}{2\pi}\sqrt{\frac{2}{L}(R_M - 1)}}$. Then we use the beam divergence

$\theta = \frac{\lambda}{\pi w_0}$. Because the numerical aperture NA is the sine of the beam divergence, as the

divergence angle is small, NA is approximated by the divergence angle. When we substitute w_0

in this equation for what we solved for above, we end up with the equation

$$NA \approx \theta \approx \sqrt{\frac{2\lambda}{L\pi}\sqrt{\frac{L}{2R_M - L}}} \, , (2)$$

where lambda is the wavelength, L is the length of the cavity, and R_m is the radius of curvature of

the mirrors. The value $2R_M - L$ is the distance between the foci of the two mirrors, directly

related to how close the mirrors are to being concentric. Based on the precision of the calipers we

used to measure the distance between the mirrors, we can reasonably assume that this value is

within 0.1 mm. Substituting our values into this equation yields an NA of around 0.036.

3.2 Finesse:

The Finesse can be predicted based on the reflectivity of the mirrors using the equation:

$$F \approx \frac{\pi}{1 - r}, (4)$$

where r is the reflectivity of our mirrors.

4. Results:

4.1 Numerical Aperture:

Using a CMOS camera (see Fig. 5) connected to a laptop (Fig. 8), we were able to estimate the

width at half-maximum of the beam. Using computer software, we created a graphic showing the relative brightnesses of the beam across a cross section of the image. This allowed us to

calculated the beam width at half-maximum. This, combined with the distance of the camera

from the center of the cavity (see Fig. 9), allowed us to estimate the NA of the cavity, and

compare it to our expected values.

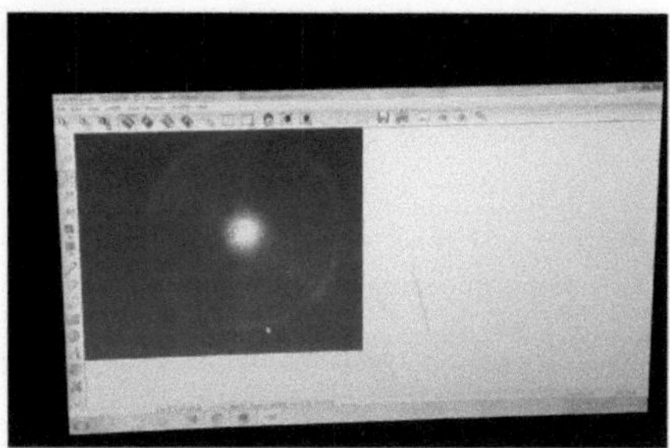

Fig. 8: A picture of a Gaussian (TEM$_{00}$) mode on the camera. This is the eigenmode of our

Fabry-Perot cavity. We used a cross section of this picture comparing the relative brightnesses to

calculate the NA for our cavity.

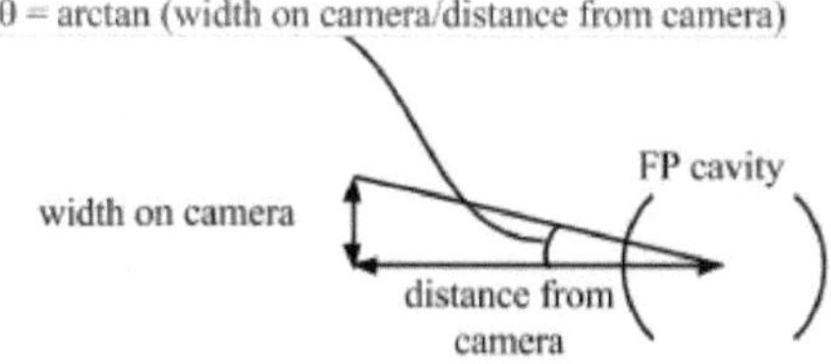

Fig. 9: The relationship between the angle of divergence with the width on the camera and the

distance from the camera

Using this method, we determined the width of our beam at half-maximum when it hits

the camera to be around 1.6 mm, with the camera set up approximately 10 cm away from the

center of the cavity. This allows us to estimate the acceptance angle of the cavity to arctan(1.6

mm/100 mm)=0.016 rad, yielding a numerical aperture of 0.016 After taking 5 measurements, we had an average value of 0.0158, with a standard deviation of 0.00284.

Based on Eq. 2, after accommodating the precision of the calipers we used to measure the distance between the mirrors, we can reasonably assume that we were able to get within 0.1 mm of this distance. Substituting the values specific to our cavity into this equation yields an NA of around 0.036, which is higher than our experimental results.

4.2 Finesse:

Using the oscilloscope and a waveform generator, we scanned the laser frequency across several cavity resonant modes, by using a TLB-6700 Tunable Diode Laser from Newport. We used a waveform generator to directly apply a sawtooth signal to the modulation port of the laser controller box. This oscillates the laser frequency across a range large enough to clearly see the eigenmodes of our cavity. A photodetector (photodetector 2 in Fig. 5) captured the light emitted by the cavity, which fed to an oscilloscope, where the eigenmodes appeared as spikes (Fig. 9).

Fig. 9: Examples of resonant modes on the oscilloscope. The spike in the middle of the screen shows the power spiking at the specific frequency. The narrow width of the line demonstrates our very low linewidth.

Using this, we were able to trigger the oscilloscope to the highest resonant mode to view

the linewidth of the largest mode. We observed our linewidth to be approximately 500 kHz

(possible error ±10%). This was measured using the cavity ring-down method, calculated using

long it took for the light in the cavity to decay to half the original power. We also approximated

the linewidth based on the width of the peak and the speed of the ramp signal. Measurement of

500 kHz was based on the spread from at least 3 independent measurements, all of which were

between 450 kHz and 500 kHz. This gave us sufficient data to approximate the actual Finesse (F)

of our cavity.

The Finesse is the ratio obtained by dividing the distance between two adjacent resonant

modes, the free spectral range, by the linewidth of the mode at half power. Thus it is represented

by the equation:

$$F = \frac{FSR}{\Gamma}, \quad (3)$$

Where FSR is the free spectral range and Γ is the linewidth of the mode. Using the

equation $\Delta v = \frac{c}{2L}$, where c is the speed of light and L is the length of the cavity, we determined

our FSR to be $\frac{c}{2(.0254 \ cm)} \approx 11.8$ GHz. This yields a finesse of $\frac{11.8 \ GHz}{.0005 \ GHz} \approx 24000$.

Based on Eq. 3, plugging in our mirror reflectivity of 99.99% yields a theoretical finesse of

$F \approx 30,000$. This agrees reasonably well with our experimental data.

4.3 Coupling Efficiency

By using the oscilloscope and a nanopositioning screw with a thread of 1 μm, we were able to

obtain a set of data that measures the coupling efficiency of our cavity as the distance between

the mirrors was changed, as shown by Figure 10.

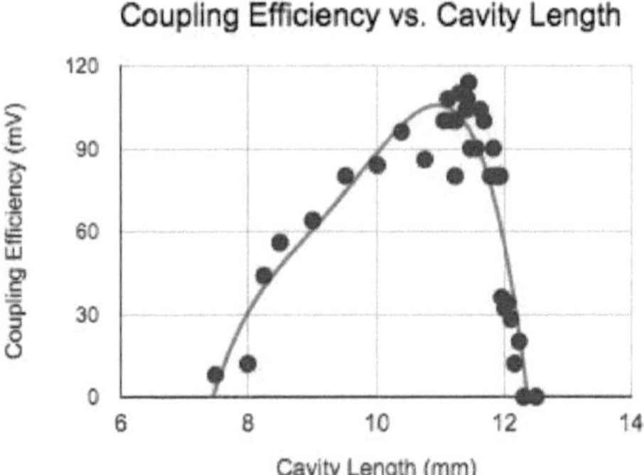

Figure 10: The graph of coupling efficiency measured in mV vs the cavity length in mm. Note that this is a minimum value - our true coupling efficiency may be higher

As this figure shows, the coupling efficiency spikes when our cavity length reaches just under 12mm. These lengths are an approximation because it was extremely difficult to precisely measure the distance between the centers of our curved mirrors, especially because obtaining such a precise measurement would require handing the fragile supermirror outside of the mount. According to the theory, the coupling efficiency should spike at the same time the numerical aperture and finesse do, at 1" between the two mirrors so that the foci of the mirrors overlap.

5. Discussion:

Our cavity was specifically designed to have a high finesse and a moderately high NA. Most previously designed cavities maximized finesse, while allowing NA to drop, approaching zero. Because we must consider the power of the laser entering the cavity, we decided to settle for a lower finesse in favor of having a higher NA. Cavities near a million finesse are within the state of the art, but this makes for a very sensitive angle of acceptance, with an NA of near zero [10].

This makes it unsuitable for our purposes, as we need for the entry beam to be of at least a certain diameter to supply the necessary power. Fiber-Fabry Perot cavities have also been manufactured using careful crafting of an optic fiber that produce higher NA and finesse [11], but these are small, delicate cavities not suitable for the wide beam and and high power required for phase contrast.

Our cavity was able to demonstrate very high finesse, but still maintained a moderately high NA. This was necessary to achieve a cavity with lower required input laser power in order to sustain the power required for phase contrast.

Using the measurements from our cavity, we can calculate the power necessary from the supplying laser, which must continuously supply power to the cavity to make up for the 0.01% of power lost. The necessary laser power can be calculated using the following equation:

$$\frac{\delta}{\pi/2} \simeq \gamma^{-1}(P/\mathrm{kW})(\lambda/\mu\mathrm{m})(v/c)^{-1}N, \quad (5)$$

where δ is the deflection of the electrons, $\gamma = 1/\sqrt{1 - \frac{v^2}{c^2}}$, v is the velocity of the electrons, P is the power required, and $N \simeq \frac{0.030(NA)}{1-r}$, where r is the reflectivity of our mirrors [6][1].

We want our phase shift in the electrons to be $\pi/2$, so $\frac{\delta}{\pi/2} = 1$. The FEI Titan electron microscope at Berkeley has electrons with energy between 80 and 300 keV, going atvelocities between 50% and 78% of the speed of light. For electrons at half the speed of light for our cavity with mirror reflectivity 99.99%, for example, the laser power required to achieve the necessary deflection at 1064 nm is P = 1.9 W. Assuming our NA is 0.1, for safety, this gives us a power requirement of 19 W. This is well within the range of values possible of an ordinary CO2 laser, so we have shown that it is practical to manufacture such a system to achieve phase contrast.

---

[1] *Note that in reference [6], the factor of NA is missing from N*

6. Conclusion and Future Work:

6.1 Future Work:

6.1.1 Numerical Aperture

The first issue that should be addressed is the unusually low NA. We suspect that the actual NA

of the cavity may be higher than our measured value. One reason for this is that Equation 4,

derived above, predicts that as the separation of the mirrors reaches the concentric case, the NA

should spike, and reach a maximum when the mirrors are infinitely close to concentric. We were

not able to observe such a trend after many attempts, so we suspect that there may be an

unaccounted factor or problem with our setup or something that otherwise interferes with the

NA. We considered couple of possibilities as to why we did not achieve this value. One

possibility is that the microscope objective we placed before the cavity has been misaligned. This

would lead to the focus of our light beam not occurring at the precise center between the mirrors,

meaning that we cannot get any light into certain angles, compromising our NA. Further tests are

required to show that this is the case.

6.1.2 Plano-Parabolic Cavity

In addition, a possible alternative to the near-spherical cavity is the plano-parabolic

cavity. Such a cavity consists of a parabolic mirror facing a planar retro-reflex mirror (see Fig.

11). The advantages of such a setup include easier mode-matching, due to the incoming laser

beam being collimated, the ability to reduce loss by using low-loss dielectric coating in the

planar mirror, and easier manufacturing than the near-spherical cavity. There would be losses

from the light spilling over the edges of the mirror, however, meaning that a higher laser power

of 25W is required [6]. However, this is still well within the range of a CO_2 laser.

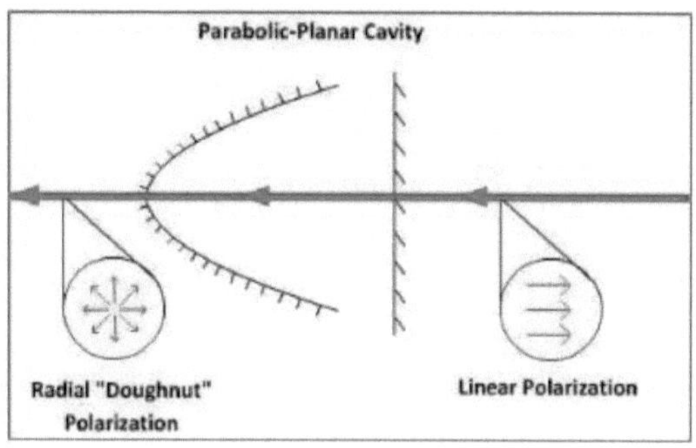

Fig. 11: Diagram for Plano-Parabolic Cavity

6.1.3 Radial Polarization:

In order to achieve the tightest focus for a plano-parabolic cavity, thereby maximizing the intensity of light, the light beam entering the cavity ought to be radially polarized. This reduces the need for laser power, and makes approaches such as the parabolic cavity much more realistic. There are several ways for making a radially polarized light beam. One possibility we discussed was using a segmented waveplate, consisting of up to 8 segments, with each segment a half-waveplate of different orientation (Fig. 12). This creates a beam with octagonal polarization. Then, the beam is focused using lenses into a very small aperture. Because only the radially polarized mode has the tightest focus, the other modes will hit the edge of the aperture, leaving the exiting beam with pure radial polarization. The beam is then collimated with a lens on the other side.

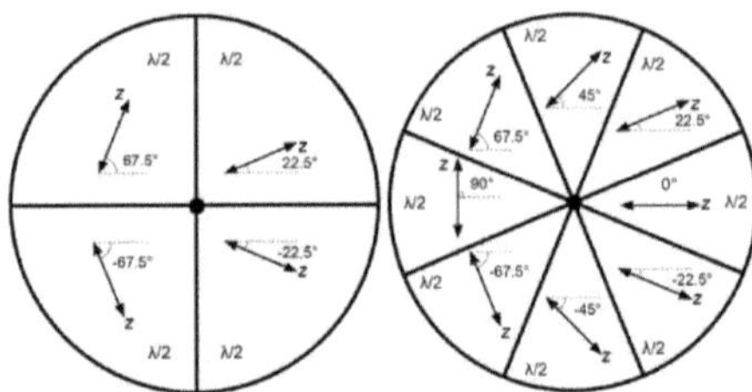

Fig. 12: Diagram showing the design of 2 segmented waveplates, with 4 segments and 8

segments, respectively [12]

Another option for creating radial polarization is commercially available "S-waveplates",

which consist of a plate of fused silica, embedded with birefringent nanogratings. This produces

similar results to the segmented waveplate [13].

6.2 Conclusion

Our results demonstrate that we have a high finesse of around 24000. This value was expectedly

high based on the high reflectivity of the supermirrors that we used.

We achieved an NA of 0.016, which was lower than our expectations. Overall, this did

not affect our results, as we still demonstrated that this led to a cavity that could successfully

achieve phase contrast. However, we suspect that the true NA of our cavity is around .036, which

would further lower the laser power necessary.

Overall, we were successful in creating a cavity that not only advances the state of the art

by creating a unique Fabry-Perot cavity with high finesse and moderate NA, but also is suitable

for work in furthering research in phase contrast electron microscopy. The specifications of the

cavity are shown to be sufficient to achieve the power necessary to effectively retard the

electrons to achieve phase contrast, making it realistic to replace carbon films.

Acknowledgements:

Our team would like to thank our mentor, Assistant Professor Holger Müller, for his incredible

generosity and patience allowing us to conduct work at his lab at UC Berkeley. We also want to

thank Michael Hohensee (now at Lawrence Livermore National Laboratory) for taking time out

of his schedule to help us in times of need. We'd also like to thank the New Focus division of

Newport for loaning us a laser controller box and laser. Finally, we'd like to thank our parents

for their support and aid over the course of this project.

References:
[1] Zernike, F. "How I Discovered Phase Contrast." 1955
<http://users.soe.ucsc.edu/~azucena/Dissertation%20Stuff/References/Zernike1955.pdf>

[2] "Phase contrast microscopy - Wikipedia, the free encyclopedia." 2006.
<http://en.wikipedia.org/wiki/Phase_contrast_microscopy>

[3] Schultheiss, K et al. "Fabrication of a Boersch phase plate for phase contrast imaging in a
transmission electron microscope." *Review of scientific instruments* 77.3 (2006): 033701.

[4] "In-focus phase contrast for electron microscopy of unstained ..." 2014.
<http://www2.lbl.gov/gulliver/presentations/Glaeser_Final.pdf>

[5] Nagayama, K. "Phase-plate electron microscopy: a novel imaging tool to ..." 2009.
<http://www.ncbi.nlm.nih.gov/pmc/articles/PMC2883085/>

[6] Mueller, H. et al."Design of an electron microscope phase plate using a ..." 2010.
<http://arxiv.org/abs/1002.4237>

[7] Mansfield, John F. "Schematic of the Optics of a Transmission Electron Microscope." n.d.
<http://www.emal.engin.umich.edu/courses/bilello460/img007.jpg>.

[8] "Introduction to Laser Frequency Stabilization." *Introduction to Laser Frequency
Stabilization.* Newport. n.d.,
<http://www.newport.com/New-Focus-Application-Note-15-Introduction-to-Las/979235/1033/c
ontent.aspx>.

[9] "Gaussian Beam - Wikipedia, the free encyclopedia." 2014.
<http://en.wikipedia.org/wiki/Gaussian_beam>

[10] Sones, Bryndol A. "The optimization and analytical characterization of super caviy mirrors for use in the single atom laser experiment" 1987. <http://dspace.mit.edu/bitstream/handle/1721.1/43362/37519942.pdf?sequence=1>

[11] Hunger, D. et al. "Fiber Fabry-Perot cavity with high finesse" 2010. <http://arxiv.org/pdf/1005.0067.pdf>

[12] "THz Waveplates" *THz Waveplates*. Tydex, <http://www.tydexoptics.com/images/products/thz_segmented_waveplates_scheme.gif>

[13] "S-waveplate (Radial Polarization Converter)" *S-waveplate (Radial Polarization Converter)*. Altechna, <http://www.altechna.com/product_details.php?id=1048z>

YOUR KNOWLEDGE HAS VALUE

- We will publish your bachelor's and
  master's thesis, essays and papers

- Your own eBook and book -
  sold worldwide in all relevant shops

- Earn money with each sale

Upload your text at www.GRIN.com
and publish for free